EXPLORING THE SCIENCE OF NATURE

The Nature and Science of
WINGS

Jane Burton and Kim Taylor

Gareth Stevens Publishing
MILWAUKEE

For a free color catalog describing Gareth Stevens Publishing's list of high-quality books and multimedia programs, call 1-800-542-2595 (USA) or 1-800-461-9120 (Canada). Gareth Stevens Publishing's Fax: (414) 225-0377. See our catalog, too, on the World Wide Web: gsinc.com

Library of Congress Cataloging-in-Publication Data

Burton, Jane.
The nature and science of wings / by Jane Burton and Kim Taylor.
p. cm. — (Exploring the science of nature)
Includes bibliographical references and index.
Summary: Examines the physical features that various insects, birds, and animals use to fly, describing their different sizes, shapes, and functioning.
ISBN 0-8368-2108-4 (lib. bdg.)
1. Wings—Juvenile literature. [1. Wings. 2. Flight.] I. Taylor, Kim. II. Title.
III. Series: Burton, Jane. Exploring the science of nature.
QL950.8.B87   1998
591.47'9—dc21   98-6136

First published in North America in 1998 by
Gareth Stevens Publishing
1555 North RiverCenter Drive, Suite 201
Milwaukee, Wisconsin 53212 USA

Printed in the United States of America

1 2 3 4 5 6 7 8 9 02 01 00 99 98

Contents

Words that appear in the glossary are printed in **boldface** type the first time they occur in the text.

African swallowtail
butterflies sip
moisture from a mud
patch. Their wings are
folded neatly upward
to be out of the way.

The Miracle of Flight

Millions of years ago, dust, raindrops, snowflakes, and lightning were the main objects that moved through the air in Earth's atmosphere. The wind may have blown plants and small creatures around occasionally, but no animals flew on their own because they did not have wings.

An animal that could leap into the air and **glide** on outstretched wings would be able to escape its Earth-bound attackers. If it could beat its wings and stay aloft — which is true flight — it was even better off. These were very good reasons for animals to develop wings. However, **evolution** is a gradual process that occurs in small steps, one at a time. Animals did not suddenly sprout wings and fly. Parts of their bodies had to change slowly to form wings. Sometimes legs evolved into wings. A pair of legs was lost to gain a pair of wings.

The evolution of true flight was difficult and complicated. It happened only a few times in the history of the natural world. Insect ancestors were first. Their two pairs of wings probably developed from flaps used for gliding. Birds lost their forelegs to make a pair of wings. Bats also lost their forelegs to make a pair of wings, and they nearly lost the use of their hind legs, as well.

Left: A canary takes off with a little jump.

Above: A peach-faced lovebird alights. Its wings spread in a last rapid flap before its feet grasp a perch. Then it folds its wings down.

Above: An African cardinal beetle spreads its wings ready to take off.

Below: A long-eared bat flutters through the air, flapping its wings.

# Lift and Drag

Top: Red maple seeds grow in pairs. Each seed has a single wing. When the seed falls from a tree, it spins. The wing produces lift.

Flight is truly amazing because a heavy object, such as a large bird, is supported by air. Instead of falling to the ground, the bird floats and even soars upward. The wonder of flight involves the special nature of wings. Wings have to be fairly flat and reasonably thin. They work only if they are traveling quickly through the air. They have to travel at a slight angle so that the front edge, called the **leading edge**, is a little higher than the rear edge, called the **trailing edge**. The angle of a wing relative to its direction of travel is called the **angle of attack**.

Right: A swallow, swooping to catch a bee, holds its wings at a slight angle to its direction of travel. This angle can be seen in the wing that is toward the camera. It has a raised leading edge.

Imagine a wing zooming through the air with a small angle of attack. Air rushing beneath the wing is squeezed together, creating **pressure**. Reduced pressure occurs on the upper side of the wing. Increased pressure beneath the wing and reduced pressure above it result in **lift**. The wing is lifted upward and can carry added weight, depending on the speed of travel.

Speed is important. The faster a wing travels, the more lift it generates but also the more it is affected by **drag**. Drag is the force that opposes movement through air. You can feel drag by holding your hand up in a strong wind. **Streamlining** reduces drag. Animals that are the most streamlined fly the fastest.

Below: A leaf-cutter bee's wings lift the bee itself and a large piece of birch leaf, too.

Below: The green woodpecker's spread wings produce plenty of lift to carry its heavy body. Spread wings, however, also result in drag.

The woodpecker reduces drag by folding its wings for a period of time — a second or so at regular intervals.

Gliders

Top: The dragonfly has four large wings. It can glide for short distances.

For an animal to glide, it needs flat surfaces somewhere on its body to provide lift. A flying gecko has flaps of skin along the sides of its body and tail and between its toes. These give the gecko enough lift to glide from high in one tree to quite low in another.

A flying squirrel has flaps of skin between its hind legs and forelegs on each side of its body. When the squirrel leaps into the air from high in a tree, it stretches out all four legs. This then pulls the flaps open.

Spurs of cartilage grow from the squirrel's wrist areas. The tough cartilage supports the leading edges of the skin flaps, which leaves the feet free for climbing.

Below: A southern flying squirrel leaps into the air from a branch high in a tree.

As the squirrel leaps, it spreads all four legs, which pulls the skin between them tight.

Flying fish have enlarged fins that form wings. Flying fish do not flap their wings, however. They just glide. The fish shoot out of the water at great speed. They spread their fins and glide freely above the waves for several hundred feet (meters). In this way, flying fish escape their **predators**. The predators remain under the water, wondering what happened to their **prey**.

Above: Flying fish rely on speed for gliding. Their wings do not flap. If they slow down, however, they can power themselves forward again by waggling the lower tip of their tail fins in the water.

A spur of cartilage, joined to the squirrel's wrist areas, supports the leading edge of the skin flaps.

Approaching the base of another tree, the squirrel brings its feet forward, ready to land.

Whirring Wings

Top: A stag beetle is one of the heaviest flying insects. Its wings and a spiral of air provide just enough lift to keep it airborne.

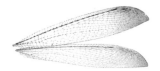

Above: The first insects to fly had four wings like these of an ant-lion.

Above: When its wings are folded, a grasshopper's hard forewings protect its delicate hind wings.

Below: A beetle's forewings, called **elytra**, are very hard. They form a tough shell around the folded hind wings.

Imagine that you suddenly became a very small flying insect. The air would feel quite different from what you are used to. It is now thick — almost like water. If you close your wings and drop to the ground, **air resistance** supports your body and you drift down slowly, landing softly. This is because you are small. Small size makes staying in the air easy. It makes progress through the air more difficult, however. Drag is much more of a problem for insects than it is for the larger animals that fly.

Drag is such a big problem for insects that they rarely glide. Large species of butterflies and dragonflies can glide, but not for long periods of time. If a small insect stops buzzing its wings and tries to glide, drag from the thick air almost immediately brings it to a halt, and it falls.

For years, scientists could not explain how insects were able to fly. More power seemed to be needed to keep a bumblebee airborne than the bee could ever produce. Recent studies have shown that there is a hidden secret to insect flight. A **spiral** current of air, or **vortex**, develops along the leading edge of an insect's wing as it flies. This vortex provides the extra lift necessary to keep even a heavy insect airborne.

Above: A bumblebee whirs its wings near a water avens flower. Bumblebees have to keep their flight muscles warm in order to fly.

11

Above: A large red damselfly climbs out of its old skin. At this point, its wings are soft, wrinkled stubs.

As the damselfly further emerges, its wings lengthen and unfold.

Insects generally have two pairs of wings. Often, the forewings are linked to the hind wings on each side of the body. The two pairs then act together like a single pair. Wasp wings, for instance, are joined by a row of little hooks on the hind wing that fit into a groove in the forewing. Butterfly and moth wings are also linked together.

Dragonflies and damselflies are different, however. Their wings are all separate. Their hind wings and forewings work out of time with each other. Flies have lost their hind wings and so only have one pair of wings. Two wings produce less drag than four, which makes flies among the fastest fliers in the insect world.

Within an hour or so, the damselfly's new wings are at full size. They are still soft at this time.

When ready for use, the new wings become stiff and clear. It has been a magical change.

Most insects buzz their wings continually in order to stay in the air. Buzzing, however, is not just flapping up and down quickly. An insect wing moves in a sort of figure-eight. It twists and moves back during the upstroke for the least amount of air resistance. The wing then comes forward for the downstroke, scooping air beneath it. This powers the insect forward and upward.

Left: As a cockchafer beetle takes off, it flaps its hind wings but hardly moves its elytra. On the downstroke, the wings are fully spread. On the upstroke, they twist and bend.

Flight Feathers

Birds are the masters of the air. They fly faster and with less effort than other fliers, and their wings fold away quickly and neatly when not in use. The birds' secret lies in their feathers. Flight feathers are fascinating structures. They are strong but very light. Each feather is **aerodynamically** shaped for its position on the bird's wing or tail. A feather has a stiff **shaft** that supports a flat **vane**. The shaft is not always central, however. On wing feathers, it is very much to one side. The shaft forms a firm leading edge to cut through the air. It is obvious whether a fallen wing feather came from the left or right side of the bird by the position of the shaft.

The vane of a feather is made of hundreds of **barbs**. These are fine, stiff threads that grow out from each side of the shaft. Along the sides of the barbs are rows of tiny hooks that engage with

Above: Newly hatched kookaburras have no feathers. Their wings are similar to arms.

In a few days, the kookaburras' feathers sprout. Each feather is protected inside a stiff tube.

hooks on neighboring barbs. In this way, a smooth, flat flight surface is constructed.

Feathers are also remarkable because they are replaced regularly. Old, worn feathers fall out, and new ones grow in their place.

Usually, birds' wings remain in good condition throughout their lives. Insects' wings sometimes wear out, and their owners can no longer fly.

Left: The barbs of a feather have many tiny hooks along each side. The hooks join the barbs together in a zipper effect.

Right: A lesser flamingo runs across the water in order to take flight. The bird needs to move quickly enough before its wings will produce enough lift for it to become airborne.

Birds' flight feathers do much more than provide flat surfaces for flight. They allow the shape of the wings to change. In flight, for instance, a bird can fold its wings slightly, causing the feathers to overlap to reduce drag.

However, a bird about to land has to slow down. To accomplish this, it needs all the wing surface it can get. A landing bird's flight feathers are well spread out to provide this maximum surface area.

Feathers also make birds efficient fliers by allowing air to pass between the feathers on the upstroke of the wing. This can be seen particularly when a bird lands or takes off. At each upstroke, the wing feathers twist so that there are gaps between them.

Above: Coming in for a landing, this little owl's wings are on the downstroke. The feathers are spread as wide as possible to give maximum lift.

On the upstroke, the situation is quite different. The little owl's wing feathers are twisted to allow air to pass between them.

Birds' wings are not just flat surfaces. Their upper sides are curved from leading edge to trailing edge. Their undersides are also curved, but less so. A **cross section** of a bird's wing is thick near its leading edge and tapers to a very thin trailing edge. This is called an **airfoil section**. At fast flying speeds, an airfoil provides extra lift without drag. It works with less angle of attack, which is what produces drag in a flat wing. Airplane wings are built in airfoil sections, but, of course, birds were using airfoils long before airplanes were imagined.

Below: The leading edges of swans' wings are thick and strong. The trailing edges are made up of the thin tips of the flight feathers. The wings show a perfect airfoil section.

Wings of Skin

Ancient animals called pterosaurs flew with wings made of skin. Some of them were the biggest flying animals that ever existed, with a **wingspan** of up to 40 feet (12 m). Pterosaurs died out millions of years ago, leaving the air to birds and insects. Birds evolved from lizard-like animals that ran on their hind legs. Bird ancestors had scales. Their overlapping scales became overlapping feathers. When small **mammals** began to fly, their fur was so different from scales that the fur could not evolve into feathers. So the first flying mammals, which were bats, grew wings similar to those of the pterosaurs.

A bat's wings are made of two thin layers of skin supported by thin bones. The bones are similar to those of human hands with four long fingers and a thumb. The thumb sticks out in front of the wing and carries a hooked claw to help the bat move around on land.

Bat flight is different from bird flight. Bats are not as streamlined and do not fly as fast as birds. Bats can turn in the air very quickly. Small bats that eat insects are able to dodge around, snatching moths and mosquitoes out of the air. Bats are agile in flight because their wing area is large compared with the weight of their bodies.

Top: A pipistrelle bat's wings fold away to little more than bones. This leaves the bat's thumbs, with their hooked claw, free for scrambling around on land.

Opposite: Pterosaurs flew on wings of skin 150 million years ago.

Below: In the air, a pipistrelle's wings provide large flight surfaces, supported by thin bones.

Hanging in the Air

Top: The kestrel is one of the largest birds that can hover. It also hangs in the air by riding upcurrents.

Some flying creatures are able to hang above the ground, as if suspended by an invisible thread. They do this by either beating their wings rapidly in order to **hover**, or by gliding on outstretched wings in **upcurrents**. Upcurrents of air form when wind meets a cliff or hill and has to go up and over the object. Birds, such as gulls and eagles, know where to find these upcurrents. They soar near the top of the windward side of rising land. Using this method, the birds travel great distances on very little energy.

Upcurrents form in another way, too. On a hot, still day, they form over flat land. Heat from the Sun warms air close to the ground. Warm air is less dense than cool air, and it rises, creating

Right: A bee-fly drinks **nectar** from a forget-me-not flower. Its wings whir at several hundred beats a second. It hovers while its feet barely touch the petals of the flower.

Left: A hummingbird flits among the flowers of a tropical tree.

strong upcurrents called **thermals**. Large birds, such as vultures, pelicans, and storks, ride these thermals, circling on outstretched wings.

Making use of upcurrents saves energy, but hovering in still air uses energy. Hawkmoths, bee-flies, and hummingbirds are all able to hover with rapidly beating wings. They get their energy for hovering from nectar. They hover as they sip the sweet liquid from flowers.

Hovering birds and insects generally have long, narrow wings. Narrow wings produce more lift with less drag than broad wings. But narrow wings have to move through the air very quickly. Hovering requires strong flight muscles as well as a great deal of energy.

Below: Ospreys hang in the air on outstretched wings, riding an up-current above a cliff.

More Uses for Wings

Top: A honeybee uses its wings as a fan to drive a current of air through the hive to cool it.

Above: A lubber grasshopper uses its wings as a warning signal, informing predators that it does not taste good.

Some flying creatures have found additional uses for their wings. Wings can be used for protection. Mother birds spread their wings to guard their chicks. Bats protect themselves while they roost by wrapping their wings around their bodies. A bird of prey may hide its kill from other birds by **mantling**, or spreading its wings around the prey.

Wings can be used for protection in another way. Some kinds of grasshoppers taste bad, and they warn predators of this by suddenly spreading their colorful wings. Butterflies and moths also

Right: A kestrel uses its wings to hide its kill from other predators.

22

protect themselves with wings. Some have bold markings on their wings that look like eyes. Showing these "eyes" scares off predators.

Some kinds of herons and storks use their wings to gather heat and also to shade themselves from the Sun while they fish in shallow water. They spread their wings to reduce the reflection of the Sun so they can see into the water.

Some bats even use their wings to help trap flying insects. While flying along, the bat uses its wings to flick an insect into a pouch formed by its curled tail. The bat then grabs the insect in its jaws for a quick snack.

Above: A marabou stork spreads its wings to gather heat from the Sun to warm its body.

Lost Wings

Top: A queen ant uses its wings for only one flight. Then, the wings break off and blow around on the ground.

Above: A male Australian bush cricket rubs its wings together to produce a continuous shrill chirp. The chirp can be heard hundreds of miles (kilometers) away. Cupped wings direct the sound.

Right: Penguins have lost all of their flight feathers. Their wings are now flippers covered in short, soft feathers.

Many kinds of creatures that once were able to fly now have weak wings or no wings at all. Extra large birds, such as ostriches, have grown much too heavy to fly. These birds use their wings only to signal to each other.

The wing feathers of cassowaries, another large ostrich-like bird, have lost their vanes — just the shafts remain. Cassowaries are deadly to attacking predators because they stab and kill their enemies with these dagger-like quills.

Penguins' wings are no good at all for flying through the air. The wings are efficient flippers, however, and penguins can "fly" through the water at great speed.

Left: An ostrich's wings carry soft, fluffy feathers and are of no use for flying. Ostriches signal to each other by waving their wings.

Insects have gone even farther in using wings for purposes other than flight. Crickets rub their wings together to produce loud chirping sounds. Other insects, such as fleas, have lost their wings completely because flying is not a part of their daily lives.

Perhaps most remarkable are ants and termites. They grow wings for one flight only, which may last for just an hour or two. When they return to the ground, they cast off their wings and start building underground nests, never to fly again.

Above: Flies, such as this crane fly, have lost their hind wings almost completely. Instead, they have two little knobs on stalks that are called **halteres**. The halteres buzz during flight and help the flies keep balanced.

25

Winging Away

Top: Monarch butterflies fly long distances.

Below: Arctic terns regularly fly the greatest distances of any bird. They spend the northern summer in the Arctic and then fly to the Antarctic for the southern summer.

Wings allow animals to travel long distances quickly and with much less effort than would be necessary for an overland journey. The clear, papery wings of hoverflies carry them across entire continents. Butterflies and moths also make long-distance flights with their scale-covered wings. Monarch butterflies in North America fly south in autumn to spend the winter resting in California and in the mountains of Mexico. Without wings, these small creatures could not travel nearly as far.

Birds cover even greater distances than insects. The swallows and swifts that spend their summers in Europe wing their way south to Africa in autumn. Swifts are perhaps the most aerodynamic of all birds. They are built for life in the air. Their

diet is entirely made up of flying insects, so they do not have to stop to feed off the ground. In fact, many spend all night as well as all day in flight. To do this, they need very efficient wings. A swift's narrow, curved wings are perfectly shaped to carry it through the air with minimum effort.

Huge wandering albatrosses spend much of their lives soaring on upcurrents over ocean waves. Their wings, too, are narrow and very long, with a span of about 13 feet (4 m).

Wings are one of nature's most wonderful inventions. They come in many different shapes and sizes, but all provide lift for their owners to carry them up and away. Humans can only marvel at these magnificent winged creatures that, of their own power, can take flight.

Above: The long wings of wandering albatrosses carry them hundreds of miles (km) over ocean waves. Albatrosses stop on land only to nest and rear their young, like these on the island of South Georgia in the Atlantic Ocean.

Below: After leaving the nest, a young swift may spend nearly a year in the air, without ever landing.

Activities:

Flights of Fancy

Birds, bats, and insects are the only animals alive today capable of true flight. By flapping their wings, they can stay in the air as long as they want. Human-made flying machines do not, as a rule, flap their wings. The action of flapping is too complicated for machinery. Airplanes use propellers or jet engines to power their flight.

Human-made gliders and hang gliders can sometimes stay in the air for hours, but they can do this only when there are thermals and upcurrents. Most gliding animals can stay up only a short while. Once in the air, they continually lose height or speed and often have to land in a hurry.

Paper Gliders

It is possible to design and build paper gliders that will fly in a very similar way to gliding animals. You will need some paper, a pencil, scissors, modeling clay, and tape. Fold a sheet of paper in half, and draw an outline of half a glider on it (*below*). Draw the outline so that the mid-line of the glider runs along the fold in the paper. The wingspan of your glider should be between

4 and 6 inches (10 and 15 centimeters). Cut the shape out, but do not cut through the fold (*below*). Some possible shapes are shown (*opposite*), but almost any shape with wings will fly.

Below: To build and fly a paper glider of your own, you will need some paper, a pencil, scissors, modeling clay, and tape.

Weigh the nose of your glider down with a small ball of modeling clay taped to it, and give it a trial launch. The best way to do this is to balance the glider on the thumb and forefinger of your loosely closed fist. Now move your fist forward and release, leaving the glider in the air with enough speed for its wings to provide lift. Several trial flights may be necessary to determine the right amount of modeling clay to use.

To fly in a straight line, your glider may need movable parts called ailerons. Fold the glider along its mid-line again and make two small cuts in the trailing edges of

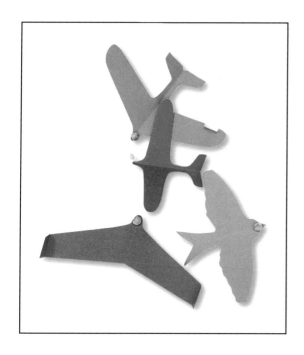

Flying Beetles

Beetle wings are thin and papery, like flies' wings. But they are kept tightly folded beneath the beetle's elytra. Remarkably, when opened, the wings are stiff enough for the beetle to fly.

Ladybugs, click beetles, and chafers are good subjects for observation. Look for them on a warm day. Be careful with large beetles because they sometimes bite, and keep away from blister beetles because they release a very nasty chemical.

Without harming it, allow a beetle to crawl over your hand and up a raised finger.

When the beetle gets to the tip of your finger, it may reach around with its legs to make sure it is at the highest point. It may then spend some time polishing its legs and antennae before taking flight.

When the beetle opens its wings, see what it does with its elytra. Most beetles spread their elytra sideways (below), but burying beetles fold their elytra upward, along their backs. Some kinds of chafers flick their wings out without opening their elytra at all.

As a flying beetle lands, watch it carefully to see an amazing vanishing trick as it stows its delicate wings away beneath its elytra. Be sure to return the beetle, unharmed, to the wild.

the wings near their tips, to form flaps about 1/2-inch wide and 1/8-inch deep (10 x 4 millimeters). If your glider has a tendency to circle to the right, bend the left aileron up slightly. If it circles to the left, bend the right aileron up slightly. Keep making adjustments to the ailerons until the glider flies straight.

Broad, thick wings generally produce a steady glider that does not get very far. A glider with long, narrow wings will fly better but may be unstable and may crash frequently. Experiment with different wing shapes. Probably the most efficient flier is a glider that is all wing. It has the most lift and the least drag. But a flying machine that is all wing is very unstable.

Skill and patience are required to successfully fly a paper glider. Evaluate your skill and your patience by measuring the distances your gliders cover.

Glossary

aerodynamic(ally): designed to produce lift and move through the air smoothly. The more aerodynamic a bird's wings are, the greater its flying efficiency.

air resistance: a force that opposes movement through air.

airfoil: the cross-sectional shape of a wing that produces lift.

angle of attack: the angle that the flat surface of a wing makes relative to its direction of travel through the air.

barbs: the fine threads that form the vane of a feather.

cross section: the appearance of an object that has been sliced across to reveal its interior.

drag: a force that opposes movement of an object through air or water.

elytra: the forewings of beetles that have hardened to form wing cases.

evolution: the gradual, orderly change from one form to another. Over millions of years, plants and animals evolve into new species.

glide: to move smoothly through the air without engine power.

halteres: the knob-like organs of flies that replaced their hind wings.

hover: to remain suspended in mid-air.

leading edge: the front edge of a wing.

lift: the force that tends to raise an object as it moves through the air.

mammal: a warm-blooded, furry animal that produces milk for its young.

mantling: when birds of prey spread their wings to hide their prey.

nectar: the sweet liquid produced by flowers to attract bees, birds, and other animals.

predator: an animal that hunts other animals for food.

pressure: a force that squeezes or compresses a substance.

prey: an animal that is hunted by other animals for food.

shaft: the stiff rib that runs through a feather.

spiral: a continuous line that curves inward or outward from a central point.

streamlining: a certain shaping that allows a body (or object) to move through air or water easily.

thermal: updrafts of warm air on land usually generated by the Sun.

trailing edge: the rear edge of a wing.

upcurrent: an upward movement of air.

vane: the flat part of a feather that is made of hundreds of barbs.

vortex: a spiraling current of air or water.

wingspan: the distance from the tip of one of a pair of wings to the tip of the other.

Plants and Animals

The common names of plants and animals vary from language to language. But plants and animals also have scientific names, based on Greek or Latin words, that are the same the world over. Each plant and animal has two scientific names. The first name is called the genus. It starts with a capital letter. The second name is the species name. It starts with a small letter.

Adélie penguin (*Pygoscelis adeliae*) — Antarctica 24-25

arctic tern (*Sterna paradisaea*) — worldwide 26

cockchafer beetle (*Melolontha melolontha*) — Europe 13, 29

common bee-fly (*Bombilius major*) — Europe 20

European swallow (*Hirundo rustica*) — Europe, Africa, North America, Asia 6

European swift (*Apus apus*) — Europe, Africa 27

four-winged flying fish (*Cypselurus heterurus*) — tropical seas 9

green woodpecker (*Picus viridis*) — Europe 7

Jamaican mango hummingbird (*Anthracothorax mango*) — Jamaica 21

kookaburra (*Dacelo gigas*) — Australia 14, 15

large red damselfly (*Pyrrhosoma nymphula*) — Europe 12, 13

lesser flamingo (*Phoenicopterus minor*) — East Africa, India 16

little owl (*Athene noctua*) — Europe, Asia 17

marabou stork (*Leptoptilos crumeniferus*) — Africa 23

moss carder bee (*Bombus muscorum*) — Europe 11

osprey (*Pandion haliaetus*) — worldwide 21

ostrich (*Struthio camelus*) — Africa 25

pipistrelle bat (*Pipistrellus pipistrellus*) — Europe 19

southern flying squirrel (*Glaucomys volans*) — North America 8, 9

swallowtail butterfly (*Graphium species*) — southern Africa 4

wandering albatross (*Diomedea exulans*) — southern oceans 27

water avens (*Geum rivale*) — Europe 11

whistling swan (*Olor columbianus*) — North America 17

Books to Read

Birds. Maurice Burton (Facts on File)
Birds. Flight. Young Scientists Concepts and Projects (series). (Gareth Stevens)
Birds: Masters of Flight. Secrets of the Animal World (series). Eulalia García (Gareth Stevens)

Flutter By, Butterfly. Nature Close-ups (series). Densey Clyne (Gareth Stevens)
Owls. Animal Families (series). Markus Kappeler (Gareth Stevens)
Wings (series). Patricia Lantier-Sampon (Gareth Stevens)

Videos and Web Sites

Videos

Butterfly and Moth. (DK Publishing)
Exploring the World of Birds.
   (Library Video)
Flight. (International Film Bureau)
Flight. (Wood Knapp Video)
Flying Birds. (AGC Educational Media)
Wings. (Discovery Channel)

Web Sites

home.earthlink.net/~edwardsjm/jme.
   index.html
www.wco.com/~ggothard/BirdsofPrey.htm
www.nasa.gov/NASA-homepage.html/
www.yahoo.com/Science/Aviation_and_
   Aeronautics/History/People/Wright_

Some web sites stay current longer than others. For further web sites, use your search engines to locate the following topics: *air pressure, airfoil section, airplanes, birds, feathers,* and *gliders*.

Index